ISBN 978-1-7320241-3-7

Author and Publisher:
Ebitari Isoun Larsen
Delta Data Services
Long Beach, California, USA
elarsen@dds-llc.com, +1 (562) 366-4774
Cover credit Ebitari Larsen through Adobe Stock
Formatting Assistance Queeneth Odimegwu, Fnu Ibrahim and Isaac Vu

Published November 2018

This page was intentionally left Blank

To my husband, Matthew,

who supports me in all my professional endeavors and inspires me to be the best person I can be.

To my son, Briggs,

who is an inspiration to me and blesses me with love, laughter, and awe.

To my family comprised of authors of many kinds

This page was intentionally left Blank

Table of Contents

Introduction

This document is a simple User Guide for your company's Customer Relationship Management (CRM). You'll find in the Guide:

- Screenshots and simple explanations for the various parts of your CRM and how many companies want their sales staff to use them.
- Some simple customizations you can do as a user to make your CRM more user-friendly for you in particular.

This book giver you a general overview of Zoho CRM. When you complete the book you will have a good understanding for the structure of Zoho CRM in both the desktop and mobile applications. With this information you will be able to navigate Zoho CRM and clearly understand that you are doing things according to how the desktop and mobile applications are structured.

As a next step after this book, use the Administrator's Guide to learn simple techniques to customize your CRM. But in this book, you will get a very clear understanding for how CRM is set up and you will be able to determine how to use CRM within the structure in which it was designed to match your company's needs.

So, for example, before you make any custom modules as an administrator, it is best to understand how the CRM works in case there are existing modules that can achieve what you want. After going through this book you will have a general introduction to Zoho CRM and have a base of knowledge to build on in more advanced functionality.

The Initial Startup Screen

When you open the Home screen on your CRM, there is a small tab along the side called 'Getting Started', you can click on this for help setting up your email and some other settings..

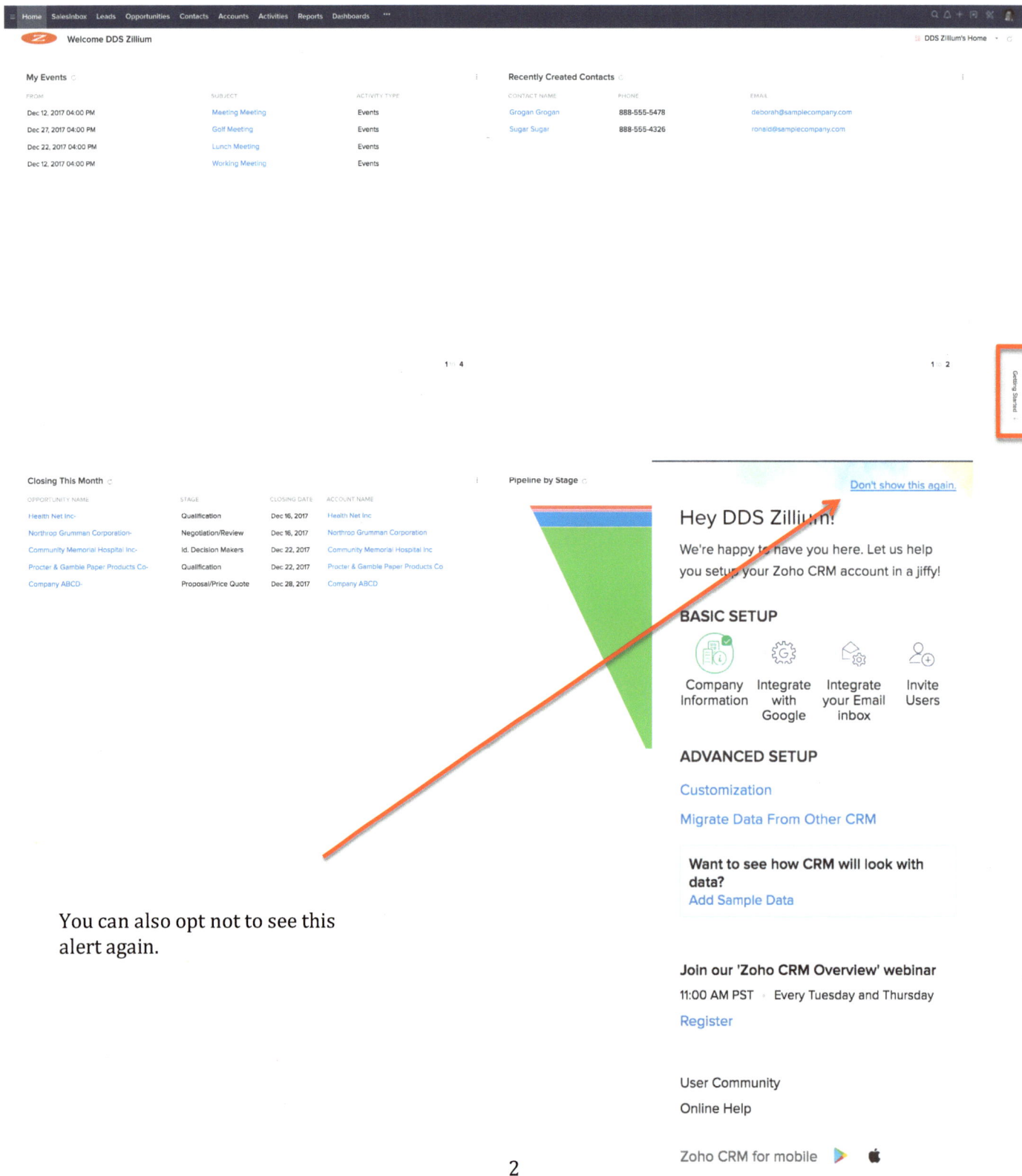

You can also opt not to see this alert again.

The Home Tab

It is recommended to select the Customized View. This view shows Views or filtered lists from the Tabs or Modules of your Choice.

Examples of Views in the Accounts module could be as follows:
- All Accounts (this would mean the organizations Accounts included those owned by all users)
- My Accounts (only the Accounts you have entered)
- Recently Added Accounts
- Accounts Added Today

Views are a way to look at a subset of your records based on any criteria that you are interested in.

This Home page is showing:
- A View from the Activities tab of 'My Events'
- A View from the Leads tab of 'Recently Created Contacts'
- A View from the Potentials tab of 'Closing This Month'
- A Dashboard from the Opportunities tab of 'Pipeline by Stage'

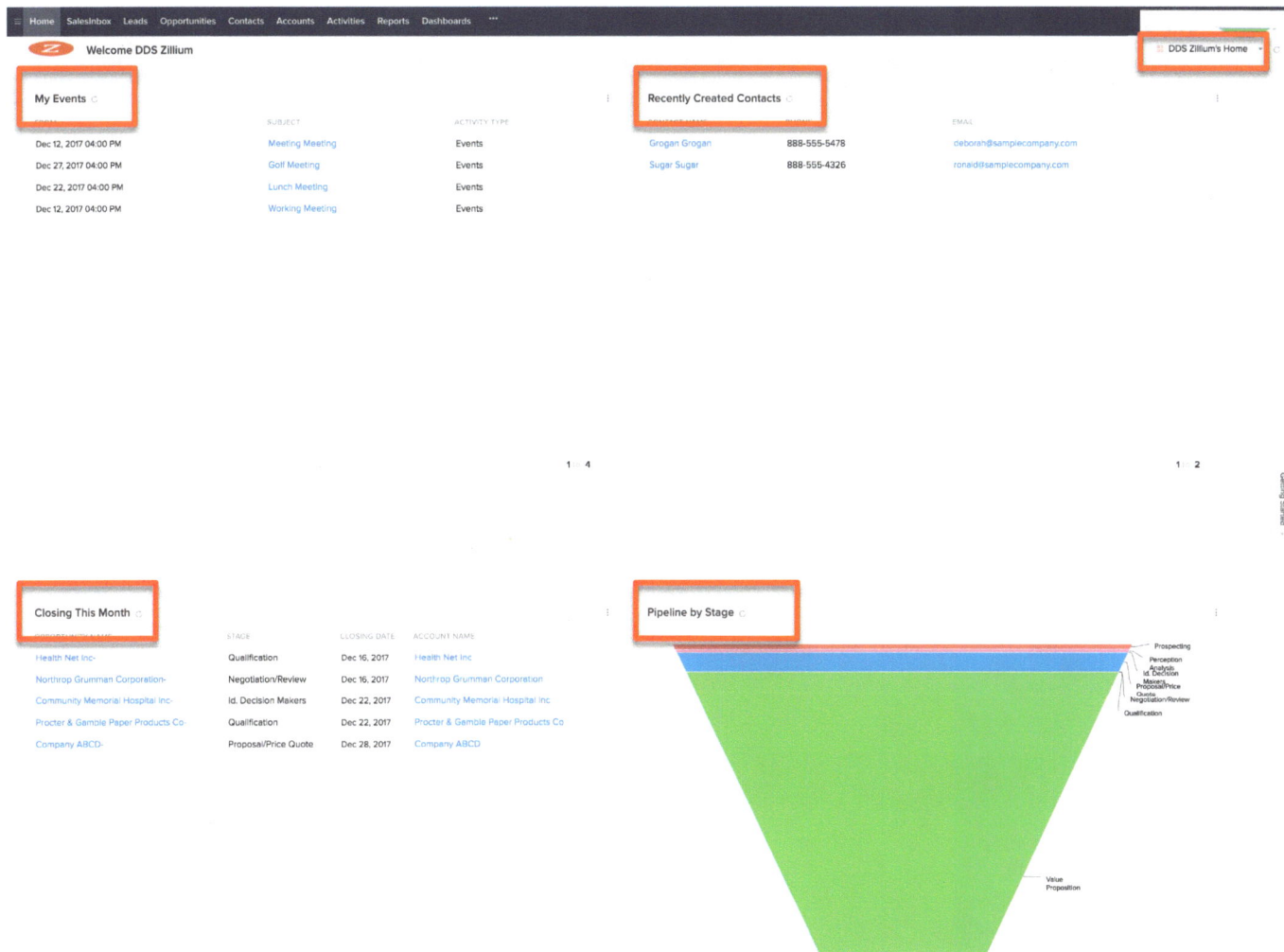

Editing the Home Tab

To change the Home page Views to ones that you want, follow the instructions below:

Edit a View or
Delete a View here

- Where it says 'Get From', select 'Custom View' or 'Dashboard'.
- Select the Module.
- Select the Custom View. There will be a dropdown list showing all the Views available to select from in that given module.
- Name the component View

Adding and Moving Components on the Home Tab

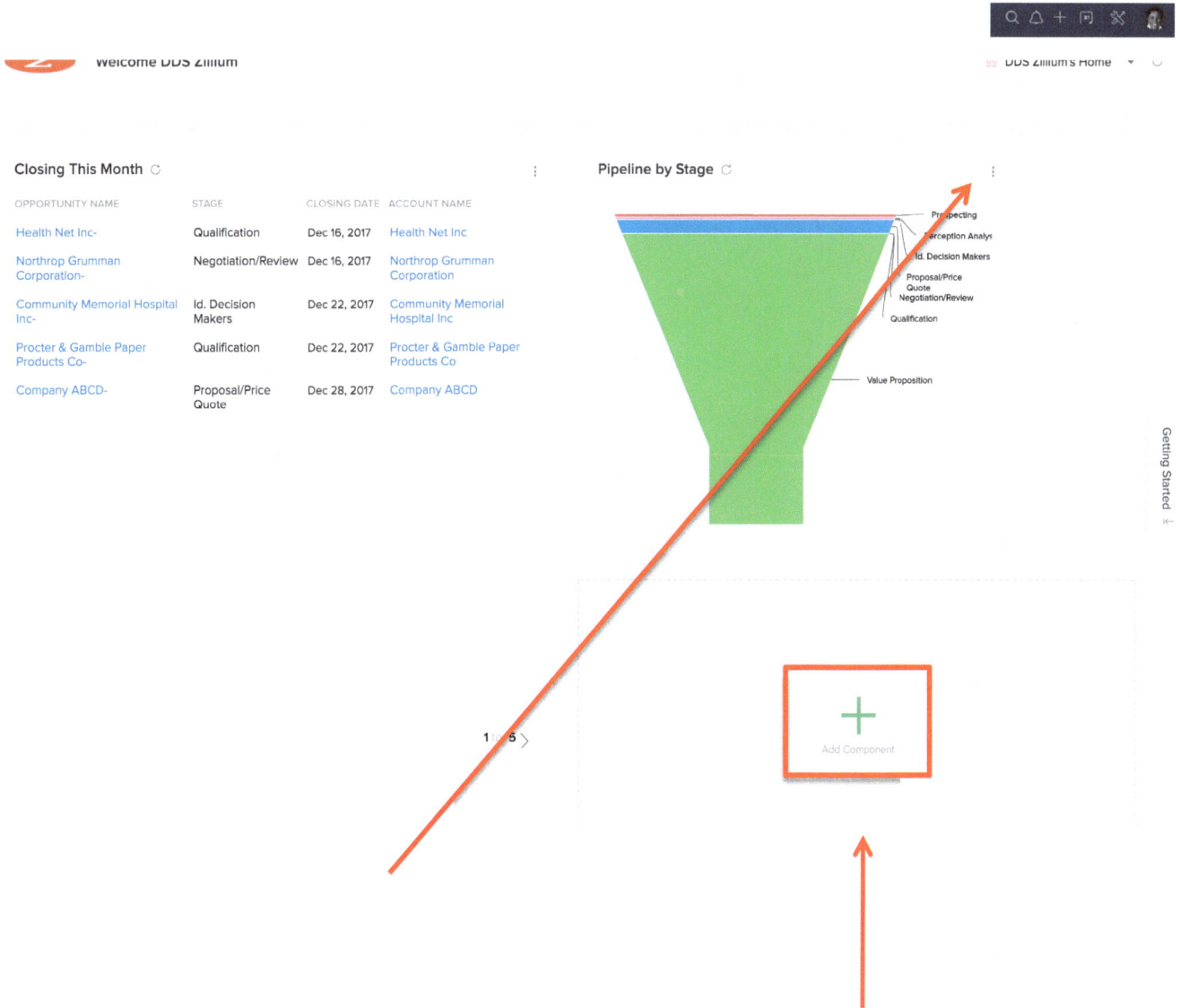

Welcome DDS Zillium

DDS Zillium's Home

Closing This Month

OPPORTUNITY NAME	STAGE	CLOSING DATE	ACCOUNT NAME
Health Net Inc-	Qualification	Dec 16, 2017	Health Net Inc
Northrop Grumman Corporation-	Negotiation/Review	Dec 16, 2017	Northrop Grumman Corporation
Community Memorial Hospital Inc-	Id. Decision Makers	Dec 22, 2017	Community Memorial Hospital Inc
Procter & Gamble Paper Products Co-	Qualification	Dec 22, 2017	Procter & Gamble Paper Products Co
Company ABCD-	Proposal/Price Quote	Dec 28, 2017	Company ABCD

Pipeline by Stage

Prospecting
Perception Analys
Id. Decision Makers
Proposal/Price Quote
Negotiation/Review
Qualification

Value Proposition

Getting Started

1 5 >

+
Add Component

To relocate a component View, hover over the upper right corner of the component View until you get a cross arrow, then drag the component View to the location you want it on the screen.

Add a View here

Ebi's Visual Guide: Zoho CRM User Guide

The Leads Tab Overview
This is what the Lead tab looks like.

This is where you Edit Views

These are a list of your Views. **Please set your default View to 'My Accounts' that way you only see the work that directly concerns you each day.**

This is where you Create Views

Search or filter by column

Click on this number to see the number of records "in that View". Note, if the number seems low, check what View you are in.

Total Count: **109**

Universal add new record

Universal search

This is where you add a new Lead or import a batch of Leads.

Click the column heading to sort ascending. Click the column heading again to sort descending.

You can select how many records to display at a given time. Mostly, we find it helpful to see the most records (100) at a time.

Import

All Leads ˅ Edit

CREATED BY ME
- Leads in Pleasanton

SHARED WITH ME
- All Leads
- Converted Leads
- Mailing Labels
- My Converted Leads
- My Leads
- Recently Created Leads
- Recently Modified Leads
- Today's Leads
- Unread Leads

+ Create View

- Created By
- Created Date
- Email Opt Out
- First Name
- Last Activity Date
- Last Name
- Lead Status
- Mobile
- Modified By
- Modified Date
- No. of Employees
- Salutation
- Secondary Email
- State
- Street
- Tag
- Title
- Zip Code

LEAD OWNER	LEAD NAME	COMPANY	PHONE	EMAIL
DDS Zilium	Dodson Dodson	Health Net Inc	888-555-4356	trevor@samplecompany.com
DDS Zilium	Gallert Gallert	Health Net Inc	888-555-5567	fred.nainavaii@samplecompany.com
DDS Zilium	Nainavaii Nainavaii	Health Net Inc	888-555-3456	fred@samplecompany.com
DDS Zilium	Wendt Wendt	Health Net Inc	888-555-6512	ed.wendt@samplecompany.com
DDS Zilium	Badaruddin Badaruddin	Conoco Phillips Corporation	888-555-6548	m.badaruddin@samplecompany.com
DDS Zilium	Malva Malva	Conoco Phillips Corporation	888-555-4701	jim.malva@samplecompany.com
DDS Zilium	Koger Koger	Presbyterian Intercommunity Hosp	888-555-7659	gary.koger@samplecompany.com
DDS Zilium	Adams Adams	Presbyterian Intercommunity Hosp	888-555-6666	daniel@samplecompany.com
DDS Zilium	Hensley Hensley	Varian Inc	888-555-8888	roger@samplecompany.com
DDS Zilium	Keefe Keefe	Varian Inc	888-555-7777	jerry@samplecompany.com
DDS Zilium	Bauer Bauer	Trona Railway Company	888-555-6666	robert.bauer@samplecompany.com
DDS Zilium	Osborne Osborne	Little Company of Mary	888-555-5555	brad@samplecompany.com
DDS Zilium	Johnson Johnson	Little Company of Mary	888-555-4444	bridgette@samplecompany.com
DDS Zilium	Ortega Ortega	Little Company of Mary	888-555-3333	danny@samplecompany.com
DDS Zilium	Wong Wong	Thousand Oaks Star	888-555-2222	victorwong@samplecompany.com
DDS Zilium	Clayton Clayton	Los Robles Regional Medical Center	888-555-1111	clayton@samplecompany.com
DDS Zilium	Anderson Anderson	Olive View U C L A Medical Center	888-555-5432	melinda.anderson@samplecompany.com

10 Records Per Page ˅   1 to **100** ›

6

Permissions in Tabs/Modules

Your administrator will set the data sharing permissions for all the modules or tabs according to your organization's policy. If a team is supposed to share and see each other's Leads but then not share Contacts after the Lead is converted, then that policy can be set up. The permissions can be set so that Leads show only for the given Lead Owner who is logged in. This Lead Owner doesn't see other Lead Owner's Leads. The View on the last page shows that all Leads showing are owned by the same owner, DDS Zillium. This is just one setting. It can also be set that any number of users see each other's Leads but don't then see the Leads for another set of users. Data sharing permissions are all customizable, according to how your organization sets it. If you believe that a different way of sharing records would be more appropriate, then speak with your supervisors and see if that way of sharing can be implemented.

Please note, the most flexible sharing permissions are available in the Enterprise Edition of Zoho CRM. In other edition, you may not be able to set up as flexible of permissions as in the Enterprise Edition.

Working with Views

This is what the window looks like when you edit or create a View.

When you edit some of the default Views, you will be unable to apply criteria to them. All custom Views you create can apply criteria.

Delete

Name the View

Specify the Criteria
For example, all Leads where Industry is 'Systems Integrator' AND Lead Status is 'Not Contacted'.

You can edit the pattern of parentheses, "and", and "or" by clicking 'Edit Pattern'.

Select the columns to display

Select whom to share the View with

Save or Clone or Cancel

Available Logical Operators for Specifying Criteria

The logical operators available for use in criteria depend on the data type of the selected field.

Field Type	Available Logical Operators
String	None Is Isn't Contains Doesn't contain Starts with Ends with Is empty Is not empty
Numeric	= <> < <= >= is empty in not empty
Date	Is Isn't Is before Is after Between Today Tomorrow Tomorrow onwards Yesterday Till yesterday Is empty Is not empty
Boolean	It true Is false Is empty Is not empty

Depending on the field type and operator used, you may need to specify more than one value in the criteria for a custom list view.

The Lead Record

The Lead Record has all the fields that make up that module. There is a way to show all the fields in the profile and a way to hide the details. If you come to a record and can't see all the details, don't worry, the Hide Details link might be clicked. Unclick the Hide Details and you will see all the information for the record. The Lead module also has a set of sub menus. All records in the various modules look similar but have slightly different sub menus according to their type.

You use the 'Convert' button to convert a Lead to a Potential, Contact and Account.

Preview information

Send Email to an individual Lead. From the list view, you can batch email to Leads or Contacts i.e. any module that is

Submenu.
You'll use Edit the most to edit the entire record.
Clone is also useful for example if you are making a bunch of Contacts at the same company with the same address.

Hide Details

Details of record

= Home SalesInbox **Leads** Opportunities Contacts Accounts Activities Reports ...				

Info

Timeline Last Update : 1165 day(s) ago

RELATED LIST +

Notes
Attachments
Open Activities
Closed Activities
Invited Events
Emails

LINKS +
What are links?

Send Email | Convert | Edit | Create Button | ...

Clone
Delete

Print Preview
Find and Merge Duplicates
Mail Merge
Meet Now!

Run Macro
Customize Business Card
Organize Lead Details
Add Related List

Dodson Dodson - Health Net Inc
♡ Add Tags

Lead Owner	DDS Zillium
Email	trevor@samplecompany.com
Phone	888-555-4356
Mobile	
Lead Status	

HIDE DETAILS ︿

Lead Information

Lead Owner	DDS Zillium	Company	Health Net Inc
Title	Manager Client Tech Svcs/CTO	Lead Name	Dodson Dodson
Phone	888-555-4356	Email	trevor@samplecompany.com
Mobile		Lead Status	
Email Opt Out		No. of Employees	0
Modified By	DDS Zillium Tue, 14 Oct 2014 01:10 PM	Created By	DDS Zillium Tue, 14 Oct 2014 01:10 PM
		Secondary Email	

Address Information   Locate Map

Street	21281 Burbank Blvd	City	Woodland Hills
State		Zip Code	91367-7073
Country	United States		

Description Information

Description

Editing a Lead or Any Record

Click the Edit button in the submenu shown in the previous section to edit the entire record. When you create or edit a Lead it looks like the screen below. Editing records in any tab or module looks similar.

Only fields in the module show up when you are editing. No related list fields show up. If you are missing fields, you may be editing the wrong module. You can know what module you are in because the module background will be highlighted in the top menu.

Fields underlined in red are required fields. If you are having difficulty saving a record, ensure that all required fields are filled in.

For the phone number, just put in the numbers, Zoho will format it for you when you save according the to the format (123) 456-7890.

If you put in formatting, yours will override Zoho's.

Editing a Single Field

You can also edit a single field at a time. Be sure to save or cancel before you can edit another single record or the entire profile.

Lead Information

Lead Owner	DDS Zillium	Company	Health Net Inc
Title	Manager Client Tech Svcs/CTO	Lead Name	Dodson Dodson
Phone	888-555-4356	Email	trevor@samplecompany.com
Mobile		Lead Status	
Email Opt Out	✓	No. of Employees	0
Modified By	DDS Zillium	Created By	DDS Zillium
	Tue. 14 Oct 2014 01:10 PM		Tue. 14 Oct 2014 01:10 PM
		Secondary Email	

Hover over the corner of a field and a pencil will appear. Click the pencil.

Lead Information

Lead Owner	DDS Zillium	Company	Health Net Inc
Title	Manager Client Tech Svcs/CTO	Lead Name	Dodson Dodson
Phone	888-555-4356	Email	trevor@samplecompany.com Save Cancel
Mobile		Lead Status	
Email Opt Out	✓	No. of Employees	0
Modified By	DDS Zillium	Created By	DDS Zillium
	Tue. 14 Oct 2014 01:10 PM		Tue. 14 Oct 2014 01:10 PM
		Secondary Email	

Some fields cannot be individually edited and will not show a pencil. You must edit the entire record to edit those fields.

Lead Information

Lead Owner	DDS Zillium	Company	Health Net Inc
Title	Manager Client Tech Svcs/CTO	Lead Name	Dodson Dodson
Phone	888-555-4356	Email	trevor@samplecompany.com
Mobile		Lead Status	-None- ▾ Save Cancel
Email Opt Out	✓	No. of Employees	-None-
Modified By	DDS Zillium	Created By	Attempted to Contact
	Tue. 14 Oct 2014 01:10 PM		Contact in Future
		Secondary Email	Contacted
			Junk Lead
			Lost Lead
			Not Contacted Locate Map

Edit the field then click 'Save' or 'Cancel' before you can edit another individual field.

Address Information

Street	21281 Burbank Blvd	City	Woodland Hills
State		Zip Code	91367-7073

Converting a Lead to a Potential

When you use the 'Convert' button, you convert a Lead to three things:

1. A Potential
2. A Contact
3. An Account

You can create A Contact and an Account without a Potential. If the Lead has the field called 'Company' filled out, then an Account will be created. If the 'Company' field is blank, then no Account will be created.

You can set where information goes during the conversion. But by default, the 'email' field goes to the Contact because you email people and Leads, and Contacts are people. You don't email Accounts so there is no functionality to email under Account.

Ensure to Check the Potential Box when you Convert

When you click the 'Convert' button be sure, in the next screen, to check the box to create a Potential otherwise, the system will not create a Potential for you. If this happens you can create a Potential manually.

If the Account exists for the Lead you are converting, Zoho will ask you if you want to add the Lead as a Contact to the existing Account or create a new Account. Make the selection that you require.

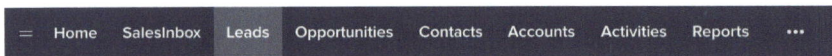

Different Ways to List the Potentials

Toggle between the two ways of looking at your list view of Potentials and other tabs like the Activities tab. See which way works best for you. We recommend the horizontal view (top) as we find it easier to work with.

The Potential Tab

On the Potentials Module, you can Mass Email, Mass Update records and mass Change Owner as well as other functions.

You may wonder how does a Potential send an email when it has no email field on it? When you send a mass email on Potentials, the email goes to the primary Contact's email address. If you send a single email, it defaults to the Primary Contact and then you can add additional Contacts to the email before you send it.

The Potential Profile and Currency

Here are some items of note in the Potential Profile view.

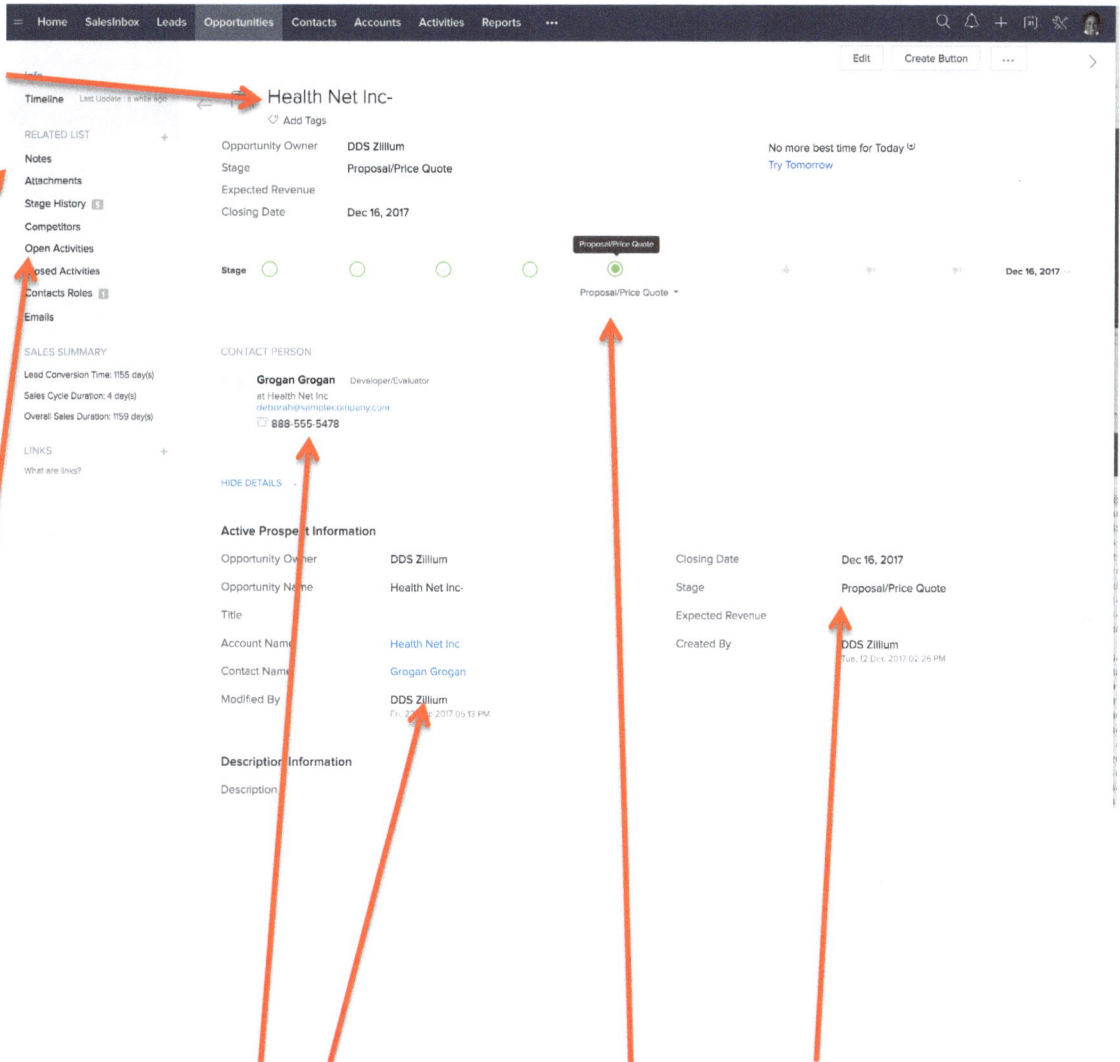

The Potential Name should be the Project name

Quick Actions:

1. Jump down to Related Lists
2. Notice that Notes are a related list. Notes in CRM are a great history as they record a time stamp when each note is taken.
3. Add new related list record like an Open Activity like a **Task**, **Call** or **Event**

This is the Primary Contact associated with the Potential.

Primary Contacts well as other additional Contacts and Accounts will also be shown in a related list if you scroll down the Potential profile.

Visual Stage History

Stage field

The Potential Stages

As you work with a Potential, change their Stage until they are 'Closed Won or 'Closed Lost'. You can change the stage in the graphic or in the field. Your organization's administrator can set the stages and probability in the Admin panel as well:

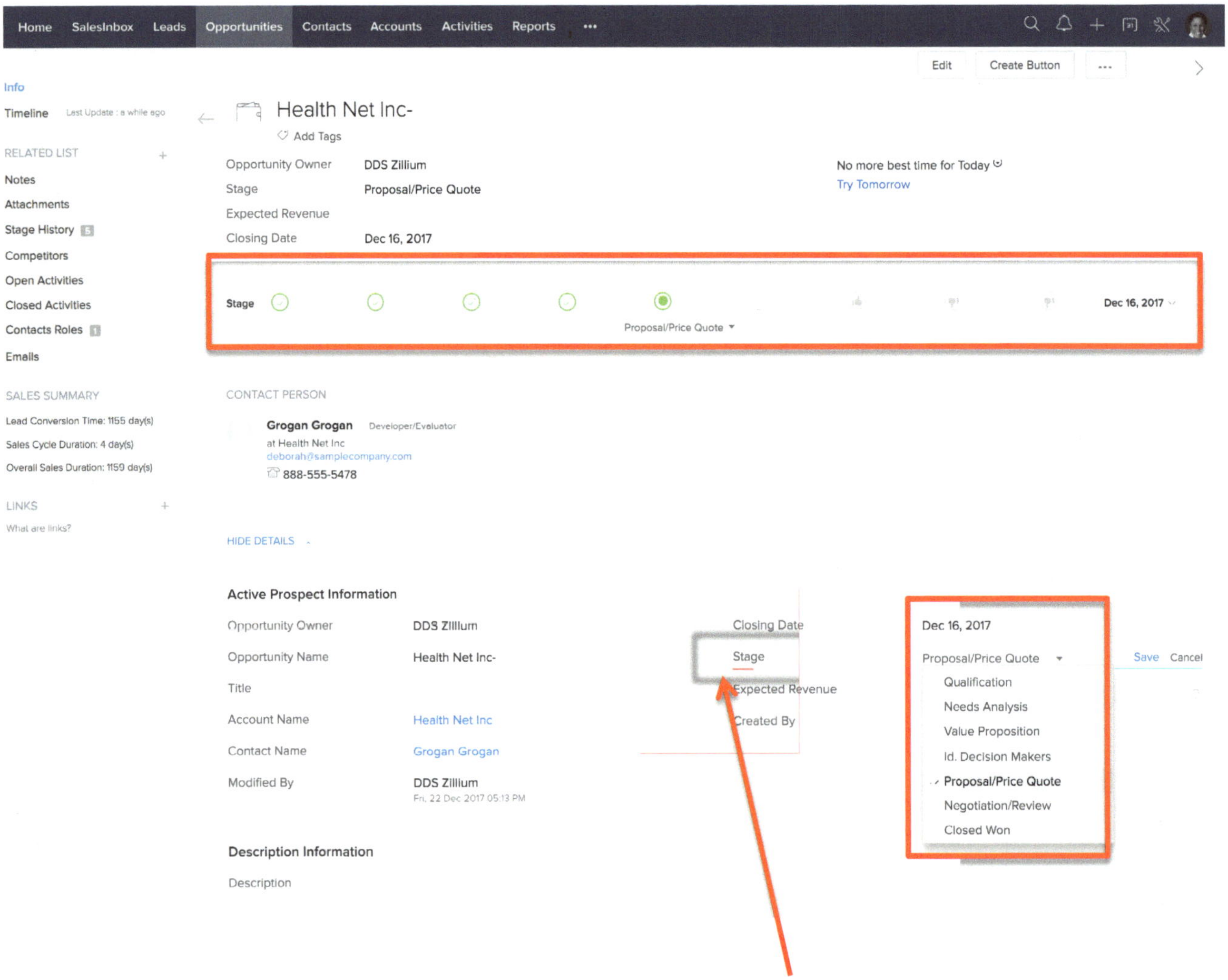

The Potential Stage History

See below how you can scroll down (also by clicking in Quick Link of the Related List on the left menu) to the content below the Potential profile and see the Stage History as you edit the Stage over time.

Click the Stage History related list link

Stage History

Other related lists

Notes related list. Note their time stamp and audit trail of who left the note.

Click the arrow to return to the top of the Potentials profile.

The Contact Tab

The Contacts tab is very similar to the other tabs in functionality. The Contact tab mainly houses the contact information for the Accounts. Consider the Accounts as the entities and the Contacts as the people or staff.

This means that a single Account can have multiple Contacts. The next section will cover how to deal with this.

The main functional difference between the Contact and Account is that you can email a Contact, but you cannot email an Account.

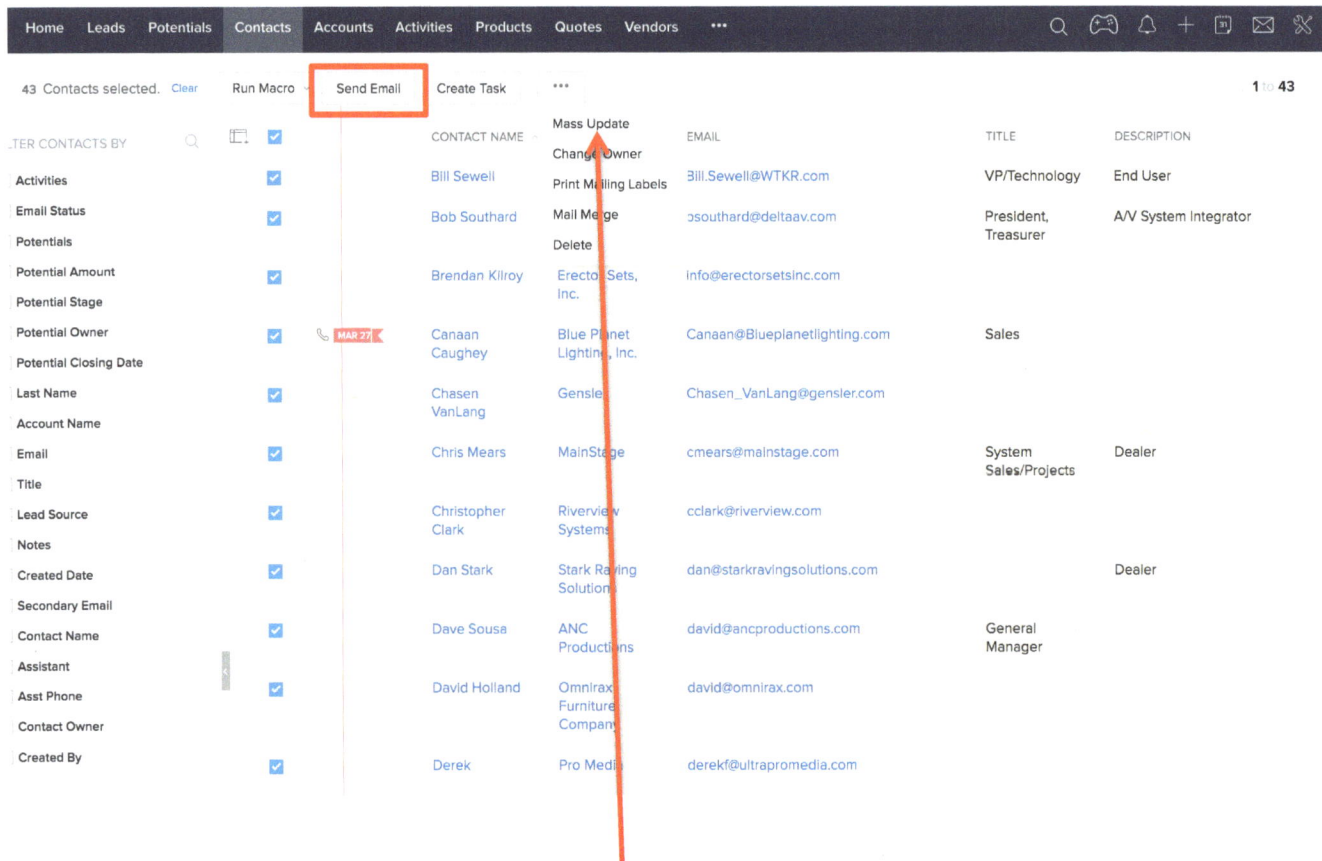

atch email Contacts.   Mass Update and Change Owner.

The Contact Profile

The emails you send Contacts (or Potentials) will show up as related lists to those Contacts (or Potentials). In the next section, it discusses related lists.

Individually email Contacts

Editing a Contact

Here is what the Contact profile looks like in edit mode. Required fields are underlined in red.

Multiple Contacts, Single Account

When you select the Account Name from the pick list in a Contact profile, it associates the Contact to that Account. From then on, scrolling down below the Account information will show the related list of Contacts and this will show all the non-Primary Contacts in that Account.

| Home | Leads | Potentials | Contacts | Accounts | Activities | Products | Quotes | Vendors | ••• |

Edit Contact

Save   Save and New   Cancel

Contact Information

Contact Owner	DDS Zillium -		Lead Source	-None- *
First Name	-None- * Culver		Last Name	Culver
Broker Name	Community Memorial Hospital Inc		Vendor Name	IT Data Processing Mgr
Email	louise@samplecompany.com		Title	
Phone	888-555-7766		Department	
Other Phone			Date of Birth	
Mobile			Email Opt Out	
Assistant			Skype ID	
Reports To			Twitter	@
Message				

Address Information   147 N Brent St   Copy Address

Mailing Street	
Mailing City	Ventura
Mailing State	
Mailing Zip	93003-2854
Mailing Country	United States

Pick the Account name from the pop-up when you click in the Account Name field.

Be sure the Account you want does not exist because typing the name in the field creates a new Account and you don't want to create a duplicate inadvertently.

Description Information

Description

End User

Save   Save and New   Cancel

+ New Account   ×

ACCOUNT NAME	CREATED BY
Loan Demo Company 1	Andrew Feldman
chrome	Andrew Feldman
Triolo Design	Andrew Feldman
Zoho	Andrew Feldman
Ebitari Larsen	Andrew Feldman
Demo Company	Andrew Feldman
Varian Inc	Andrew Feldman
Health Net Inc	Andrew Feldman
Community Memorial Hospital Inc	Andrew Feldman
Fearless Vision Project	

Selected Account: Tribune Media

This is what the pop up looks like when you click the lookup on Account.

Use the search functionality to narrow down the options and find the record you are searching for.

You have to click the magnifying glass for the search to run.

Contacts as Related List to Account

This is what it looks like when you scroll down past the Account profile in the Account and see all the related Contacts. This is called a Related List. Keep scrolling and you will see other related lists like Emails, Open and Closed Activities, Products etc.

Accoun
Name

Contac
are a
Relatec

RELATED LIST
Notes
Attachments
Potentials
Contacts
Emails
Open Activities
Closed Activities
Products
Quotes
Member Accounts

LINKS
What are links?

Timeline Last Update 2 day(s) ago

+ New

Phone	Mobile
417-332-1313	
417-332-1313	
417-332-1313	

Emails

All Contacts - Sent emails from CRM ⌄

From/To	Subject	Date	Sent By	Source	Status
usa@roevisual.com	Roe Visual at InfoComm 2017 Orlando, FL	May 11	Telemarketing	Multiple	Delivered
usa@roevisual.com	Roe Visual at InfoComm 2017 Orlando, FL	May 11	Telemarketing	Multiple	Opened
usa@roevisual.com	Roe Visual at InfoComm 2017 Orlando, Fl	May 11	Telemarketing	Multiple	Opened
usa@roevisual.com	NAB show Las Vegas	Apr 17	Telemarketing	Multiple	Delivered
usa@roevisual.com	NAB show Las Vegas	Apr 17	Telemarketing	Multiple	Opened
usa@roevisual.com	NAB show Las Vegas	Apr 17	Telemarketing	Multiple	Opened
usa@roevisual.com	Roe Visual Display Panels	Apr 05	Telemarketing	Individual	Opened
usa@roevisual.com	Roe Visual Display Panels	Mar 22	Telemarketing	Individual	Opened
usa@roevisual.com	Roe Visual LED Display Panels	Mar 16	Telemarketing	Individual	Opened
usa@roevisual.com	LDI Show, Las Vegas	Oct 05 2016	Telemarketing	Multiple	Delivered

1 to 10 >

Open Activities

+ New Task   + New Event   + New Call

Subject	Activity Type	Status	Due Date	From	To	Call Start Time	Activity Owner	Modified Time
close	Calls					Mar 27, 2017 10:19 AM	Andrew Feldman	May 30, 2017 01:58 PM
close	Calls					Mar 27, 2017 10:17 AM	Andrew Feldman	May 30, 2017 01:58 PM
close	Calls					Mar 20, 2017 12:55 PM	Andrew Feldman	May 30, 2017 01:58 PM

Closed Activities

Products

To quickly get to the related list y
want, click on the list in the left h
side of the screen. This wills scro
down.

Add a new, non-primary
Contact by clicking the
"+" sign in either location

to the i
the 'up arrow.

Creating Activities (Tasks, Events and Calls) from the Contacts Tab

The fastest way to create an Activity is from the Quick Links in any tab. It is always best to make related records from within the related record, as this will populate the related record with the related record information for you. Your supervisors want you use all three of the types of Activities and associate them to Leads, Potentials, Contacts and Accounts.

There are three types of Activities:
- Events – have a start and end time. They are for
  - Appointments – physical meeting somewhere, dinner, demonstration
- Calls – These note the time of the call. They are for
  - Outbound calls
  - Inbound calls
- Tasks – have a due date and status. These are for:
  - Creating a reminder to set an event. Tasks turn into Events.

Use all three Activity types according to the directions shown below.
- Create a new Task, Event or Call with the Related List action '+ New Task/Event/Call'

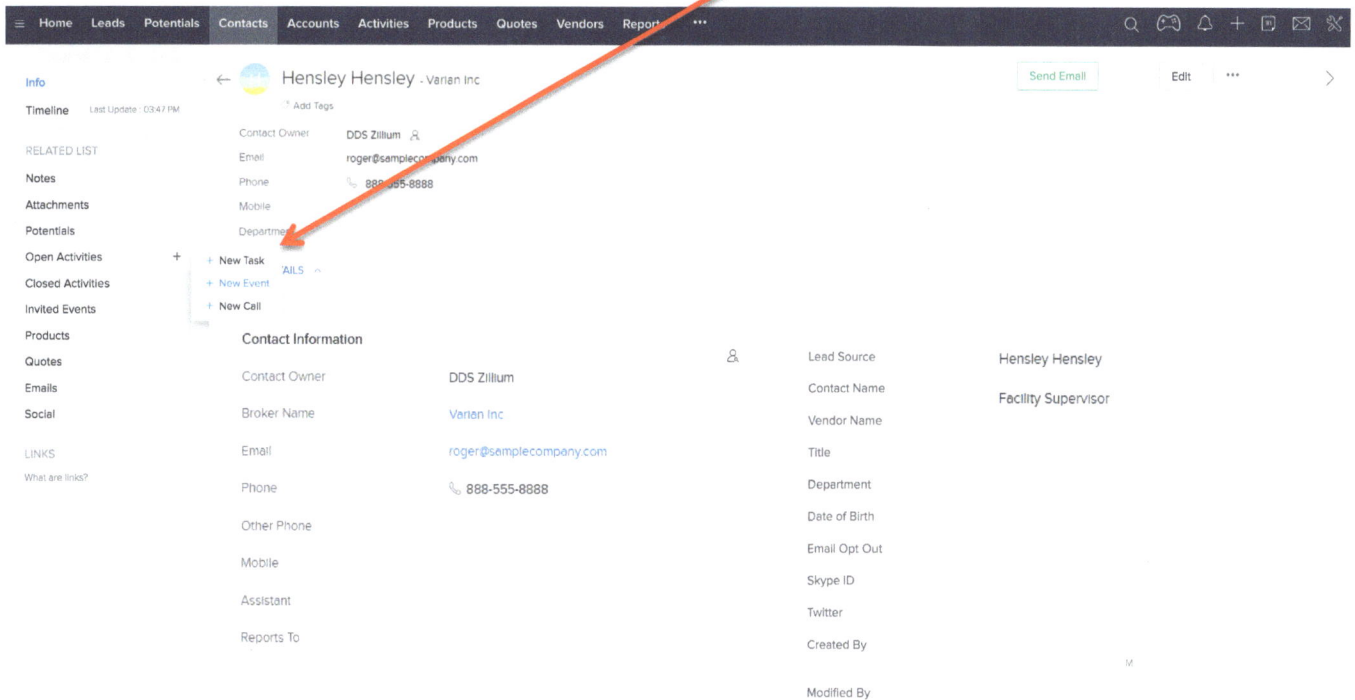

Creating a Task

It is best to make the Tasks, Events and Calls from within the Lead, Potential, Contact or Account and not from the Activities module because when you make the Activities from the module you want it related to, CRM auto populates fields like the Lead, Contact and Account Names for you and you do not have to search for them.

Ensure to enter good notes on your Activities. If you cannot save, ensure all required fields, including Description, are filled in.

This is what the Task profile looks like:

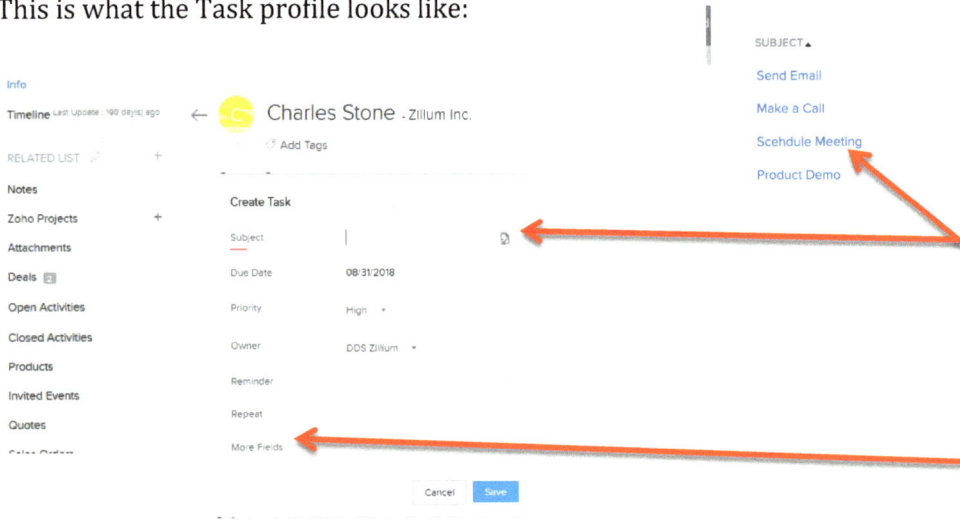

Click this icon on the subject of a Task and you get a list of options to use as your task subject.

Ensure to click where it says 'More Fields' to see the Contact and Account are populated and to add a description, set the status and send a notification email.

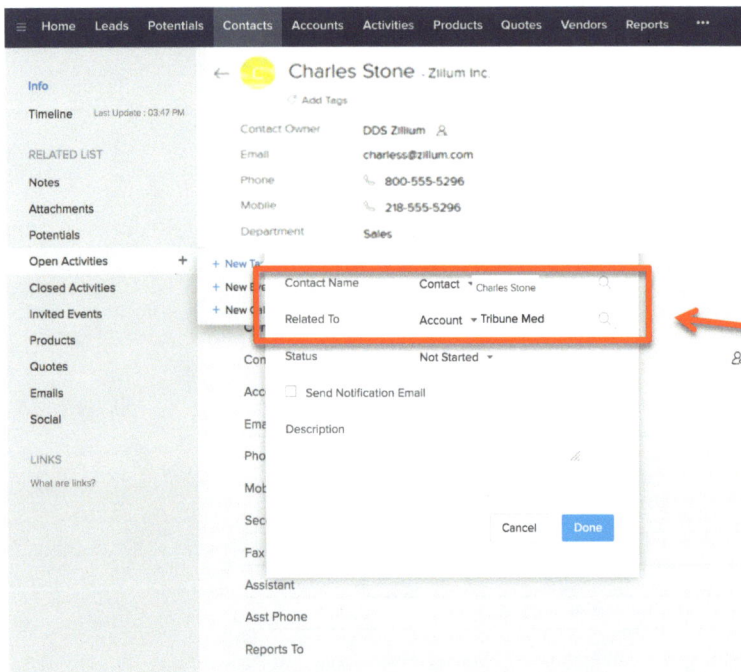

Note the Contact and Account are auto populated by creating the Call from the Account module.

Creating an Event

This is what an Event looks like in edit mode.

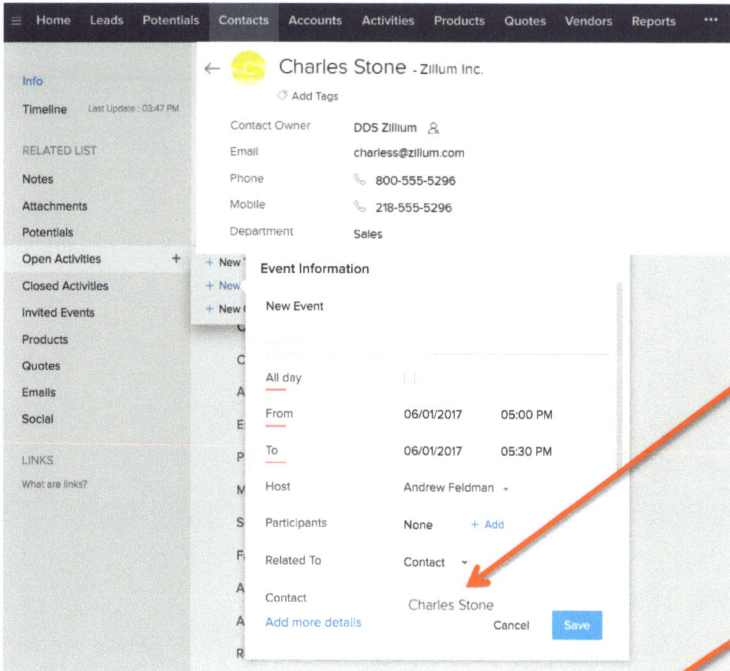

Note, you have to click 'More Details' to see the Description and Contact and Account are populated.

You can add Participants to the Event and they will be sent an invitation to add the Event to their calendar.

Contact and Account are auto populated if you create the Event from inside the Contact.

Event Information

Participants	None	+ Add
Related To	Contact ▾	
Contact	Charles Stone	
Account ▾	Zillum Inc.	
Repeat	None	
Description		
Reminder	None ▾	

Cancel   Save

Creating a Call

This is what a Call looks like in Edit mode.

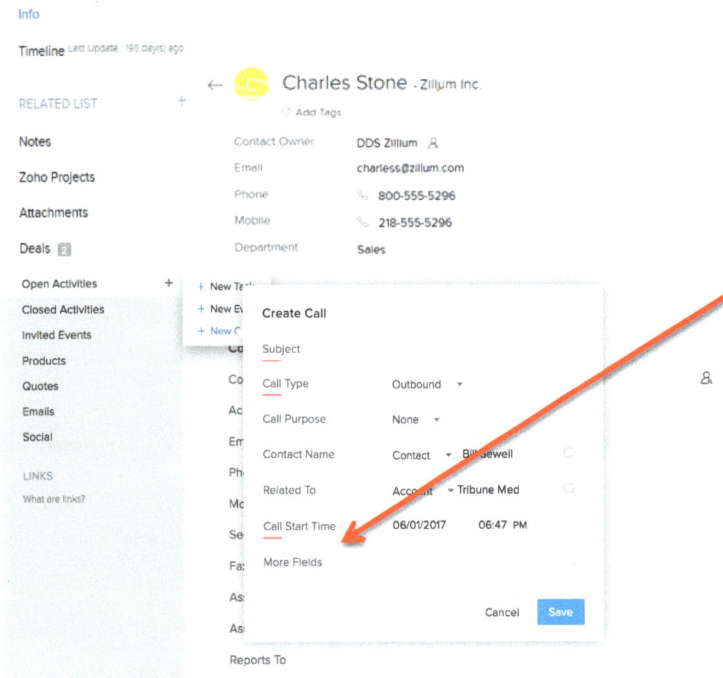

Note, the Contact and Account are populated if you create the Event from inside the Contact.

You have to click 'More Fields to see the Description and

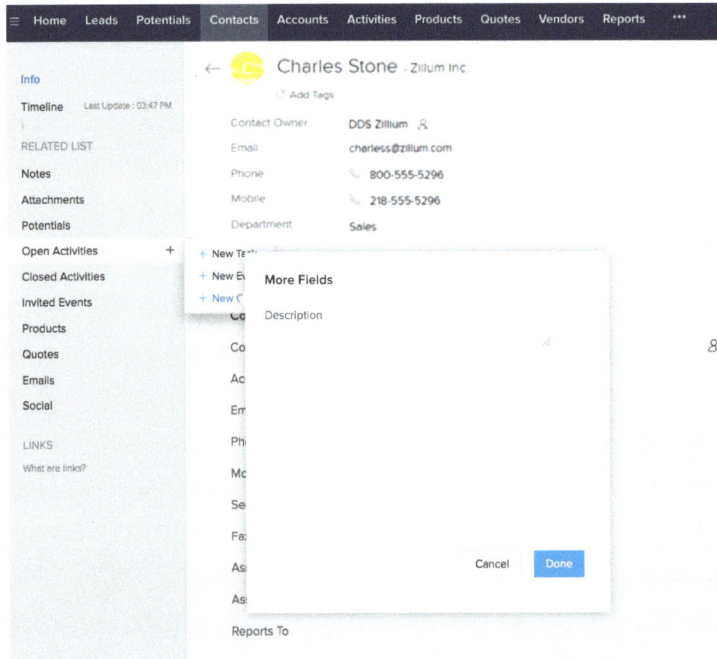

How Tasks Look as Related List

To see all the Tasks, Events and Calls associated to a Lead, contact, Potential or Account click on the left where it says Open Activities. This will automatically scroll you down. Then when you are complete, click the 'Up' arrow to return to the top of the record.

Add a new, Activity (Call or Event) by clicking the "+" sign in either location

Return to the top of the Account profile.

Emails as a related list.

Activities as a related list

The Activities Module

There are two ways to view the list of Activities. We recommend the horizontal list view, as it is more intuitive to follow. Try both types of lists and see which one works for you.

Accounts Module

The Account module is like many of the other modules, but you cannot email an account. The Accounts module, as well as Lead and Contact have a deduplication functionality for administrator roles. If your records need to be deduplicated, speak to a CRM administrator about doing it. The deduplication takes place for Contacts based on email and name and for Account based on name and phone number.

The Account module looks very similar to other modules and has similar functionality. It also has an import button. Click on the import button to import Accounts (or in any module to import records into that module).

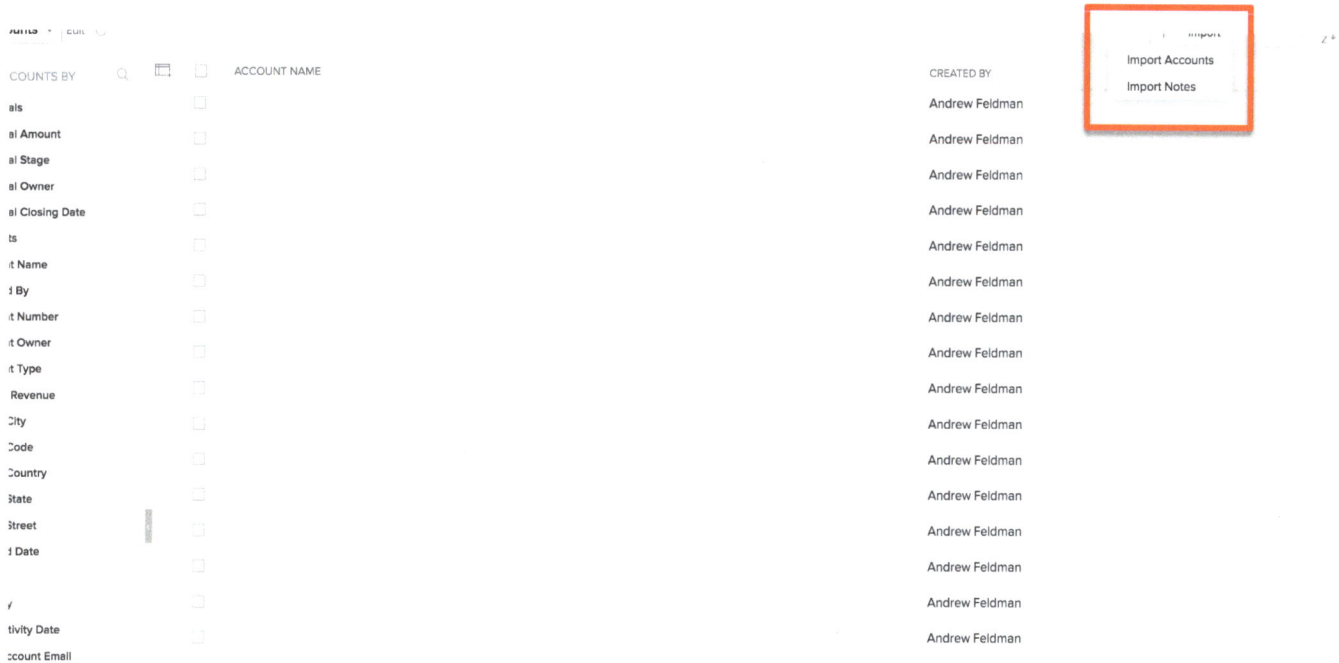

Importing Accounts (or Any Batch of Records)

If you are just importing your records, then select 'Import My Accounts'. You will not need to have a column for Record Owner with this option. If you want to assign Record Owners to several individuals, then you want to select the second option to 'Import My Organization Accounts'. Put the Owner's email address in the field for Owner.

Browse to select the Excel, .csv or other type of file that holds the data you want to upload.

Select if you want to Skip, Overwrite, or Clone duplicates and to find Duplicates based on what field. The options include email for Lead and Contact and Name for Account and Potential as well as record ID for all (but this is not something you will have accessible).

Click Next.

Importing – Matching Fields

Next you want to match the columns in your data file to the columns in CRM. This is called field mapping. Ensure that all the necessary (and required) fields are available. Ensure your data is the in the correct format for all fields. For example, don't include the currency symbol in a currency field, just the decimal number. Zoho will apply the currency symbol according to specifications in the CRM.

Click the Next button.

Importing – Finish the Import

Confirm that all fields are mapped and then click Import.

You will get a report of how many records were successfully uploaded.

If you make a mistake you can undo the import so don't navigate away from the final page until you are sure that everything is imported correctly.

If you navigate away from the page and want to undo the import, contact your CRM administrator to undo the import and then you can redo it when you have everything ready again. You will need to tell the CRM administrator the name of the file and the date on which you imported the file.

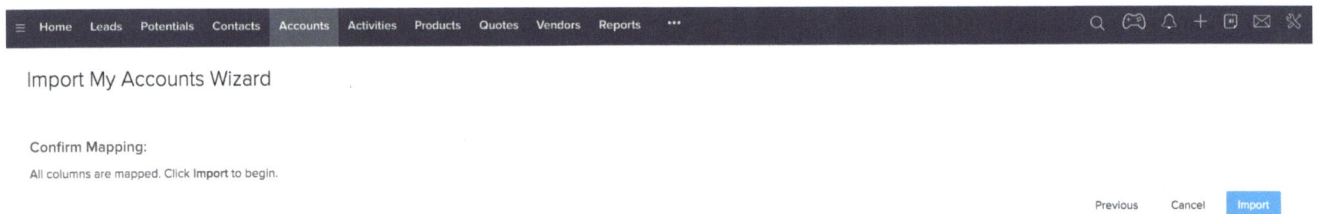

Account Profile

The account profile shows the Contact(s) as a preview at the top of the profile.

Account Edit Mode

This is an example of an Account record in edit mode. The items underlined in red are required fields.

Reports

There are various reports that are already set up as default in Zoho. Your supervisors may be requesting more reports to track the items they are interested in such as the amount of Activity on Leads and things like that. You can also make your own reports. They are created in a similar way to views but are more flexible because you can combine more than one module together. The reports will become most useful when your supervisors create the reports, they are interested in. You can track those reports and even schedule them to be delivered to you via email. You can keep track of your progress on your supervisor's reports.

You can edit the folders to show or hide which folders you want to see or not to see. Click the Edit button.

Try creating your own reports. They work very similar to the Views that were reviewed earlier in the document.

Grouped reports are very useful as you can get a count of the number or records based on up to 3 columns of your choice.

Each tab or module has a folder' for reports.

At the top of the Reports tab is a folder for 'Recently Accessed Reports'.

Reports Scheduler

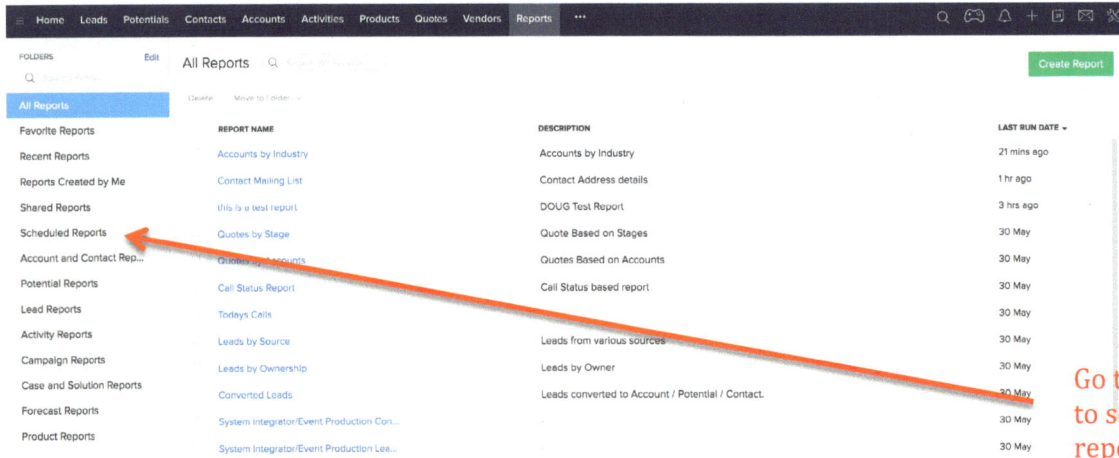

Go to this folder to schedule reports to be emailed to you on a regular schedule

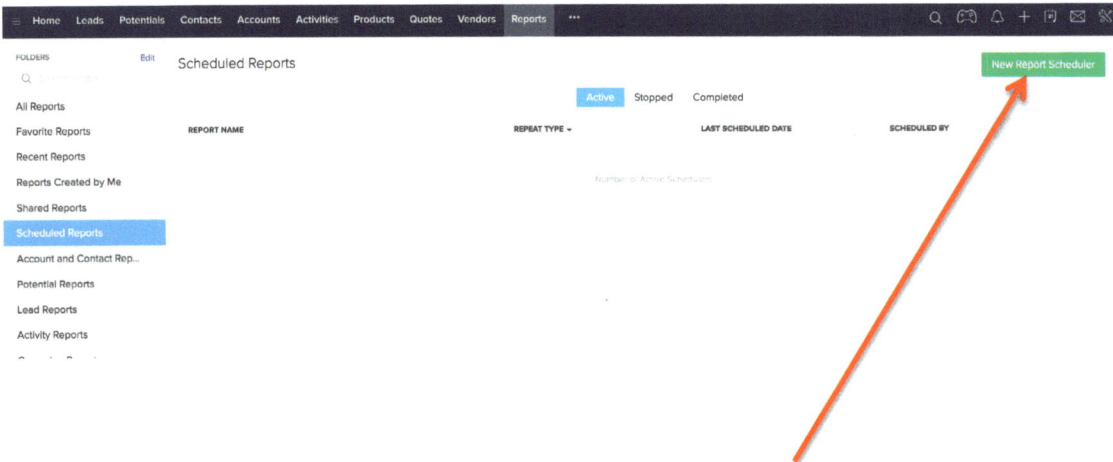

Set up a new Report Schedule

Opening a Report

Open a report. You can export it to Excel, CSV or PDF.
You can edit the report by clicking the Edit button.

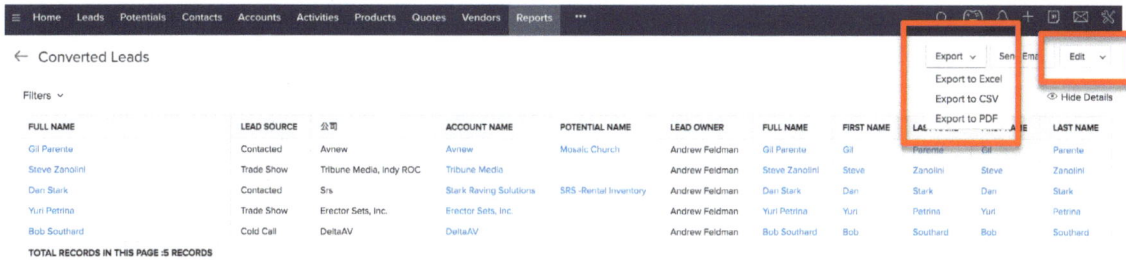

You can send the report by email immediately or schedule for later.

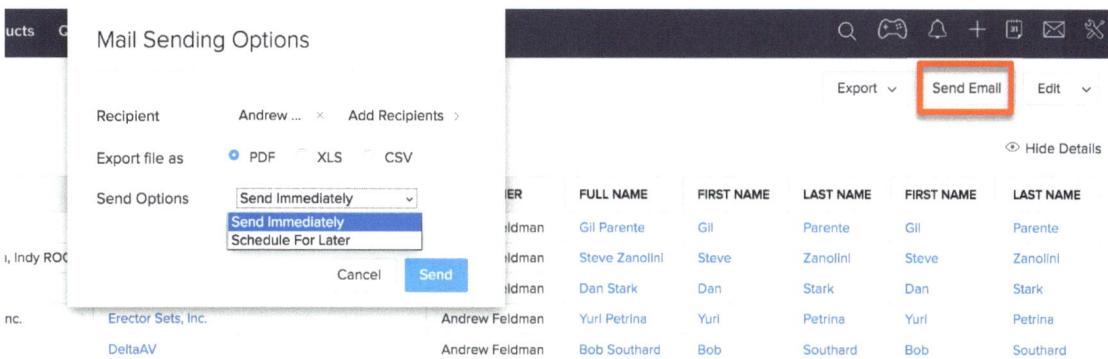

You can filter the report based on any date in the report modules.

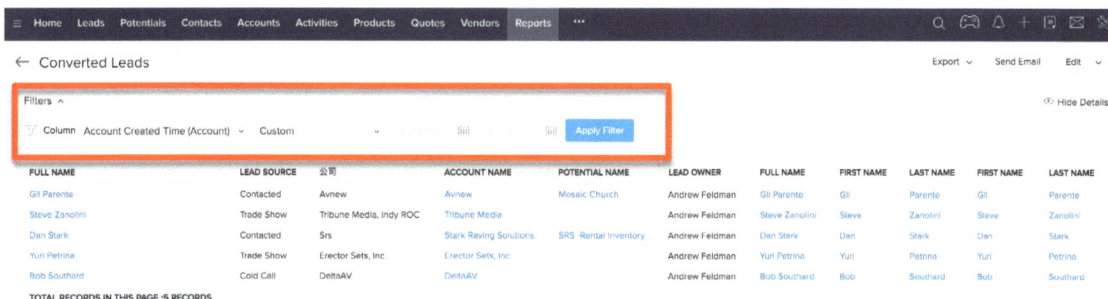

Other Modules of Interest

Other modules of interest include:

Dashboard

Add graphs from the dashboard to your home page. Follow statistics you want to track. Like the Reports tab, you will only see data that has been shared with you. If you only see your own data, then you will only see your own numbers in the dashboard. If you see others' data, then all the other data will be combined with yours in the dashboard reports.

Products

Products are the items your company sells. You need to have Products if you are going to make Quotes. This is why the Products module is enabled so that when it is time to make Quotes, the Products are ready to go. Right now, the Products have not been set up for your company. The Products will also be able to be sold in different currencies include CNY, USD and EURO. This is all to come in your CRM.

Quotes

Quotes are just what they seem. They are a way to give the customer a quote for work to be done. When the custom accepts the quote, they Quote can be converted to an Invoice. The Invoice module if hidden for now but it can be added at a time when your supervisors may want you to convert Quotes to Invoices.

Vendors

This module may be used in future to track Vendors. It is not set up at the moment.

Documents

On page 18, you can see that Attachments have been added to a Potential profile. You can add attachments from your desktop or even Google Drive. There is also a Documents module in CRM. It works like a Document Management System. You can add documents from there as attachments. However, the strength in the Documents module is for collaboration and document control. You can check out a document and work on it and then check it back in with a new version. The documents module is good as a repository for all business documents that are branded and versioned. Users can be sure they are always emailing or uploading the most recent version of a document.

The Zoho CRM Mobile App

Go to the app store on your mobile device and search for the Zoho CRM app. Download the app to your mobile device.

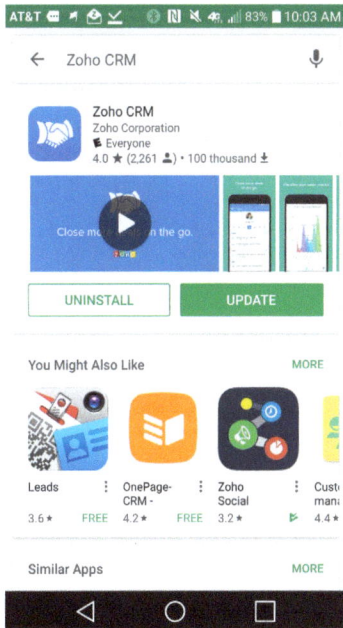

Find the app on your mobile device and open it.

Logging in to the Mobile Zoho CRM App

Log into the mobile Zoho CRM app using your Zoho (or Google) password.

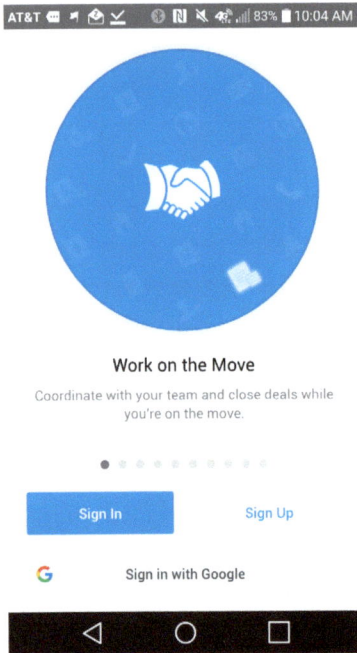

The first screen you see is the calendar view of any activities you have set up.

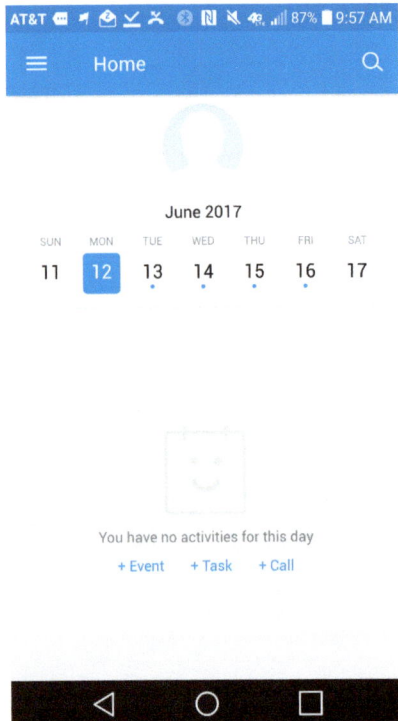

The Mobile App Menu

Find your menu for the mobile Zoho CRM app under the three horizontals lines in the upper left corner of the app.

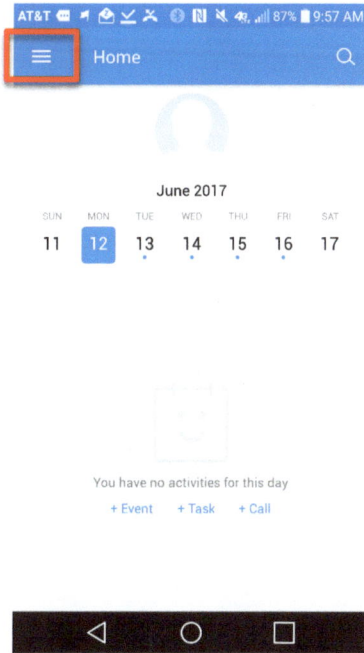

The menu will appear on the left side of the mobile app screen.

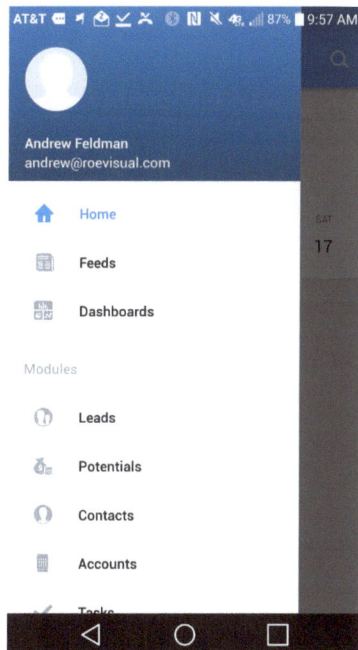

Views in the Zoho CRM Mobile App

Open the any module, like the Leads module, and ensure you are in the View that you want to be in.

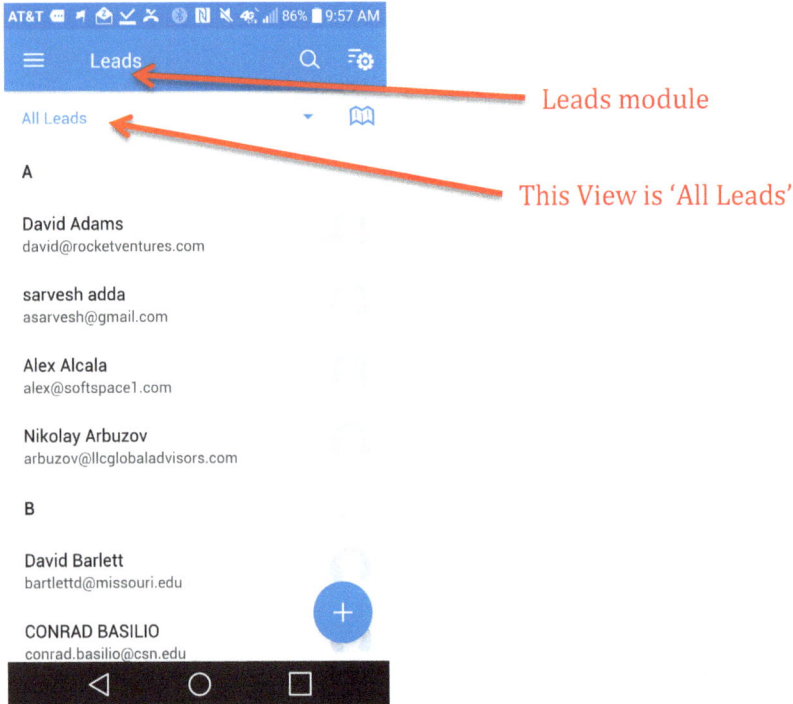

You can choose to use any View that is available.

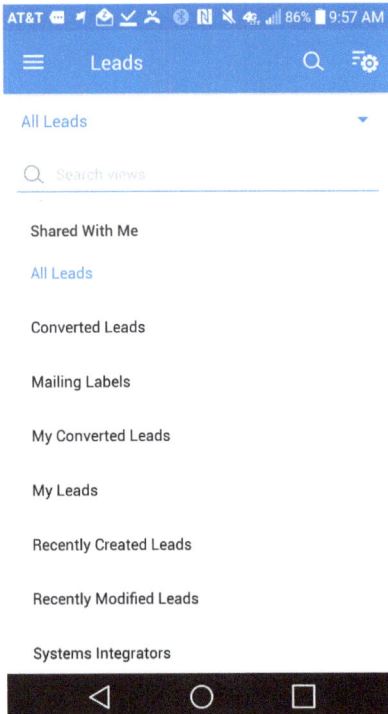

Adding a Record in Zoho CRM Mobile App

Click on the plus sign in a blue circle in the lower right corner of each module.

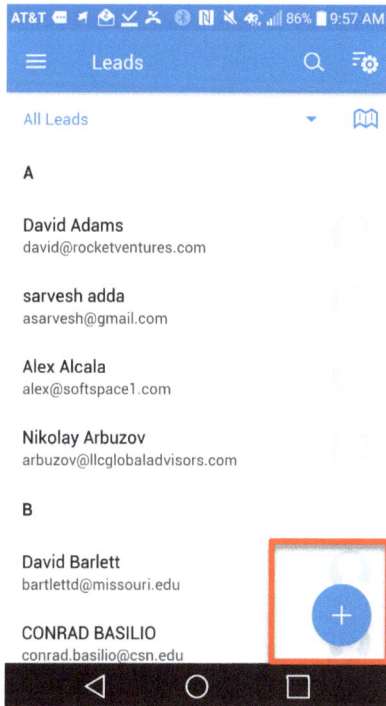

Searching for a Record in Zoho CRM Mobile App

Use the magnifying glass to search for a record Zoho CRM in the mobile app.

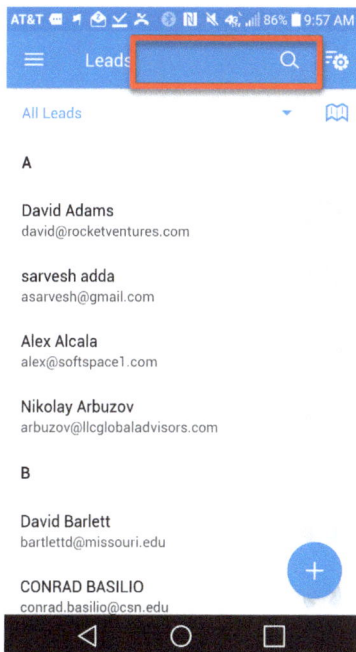

Viewing Records in the Zoho CRM Mobile App

When you view a record on the mobile app, you are first taken to a profile showing the related lists to the record. For Leads, the related lists are Notes, Attachments, Products, Tasks, and Events etc.

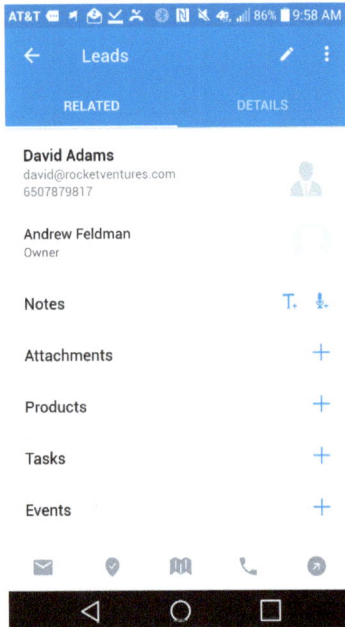

You can also view the Details, which defaults to only showing fields with data in them. You can change this type of profile view in settings and this is shown later in this User Guide.

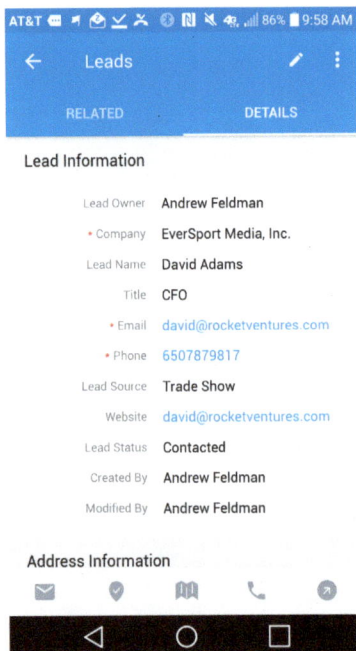

Show All Fields in Zoho CRM Mobile App

When you scroll down to the bottom of the Details view on a record, you see a link to 'Show All Fields'. Click on this to see the fields that don't have any values in them in addition to the fields that are showing by default that do have values in them.

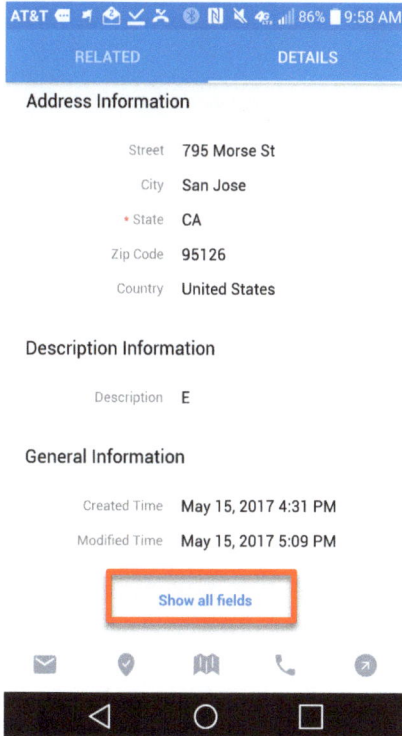

This profile view shows all fields even those without values in them.

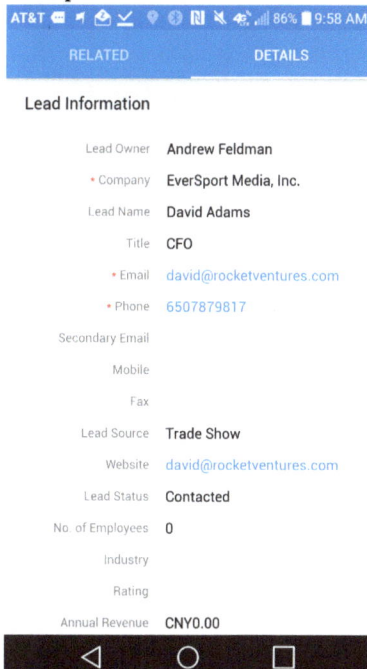

Editing a Record in Zoho CRM Mobile App

Click the pencil symbol to edit a record.

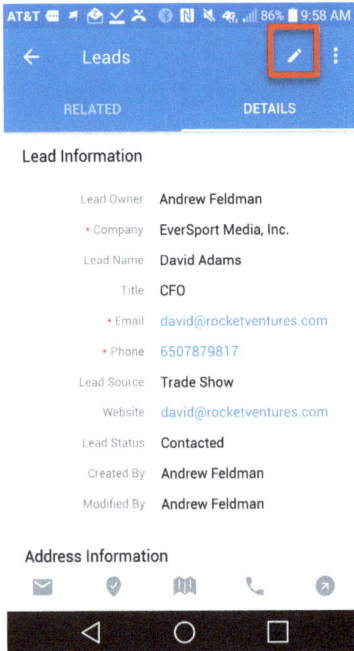

Making a Call or Sending an Email from a Record in Zoho CRM Mobile App

Open a record profile and click the email or phone icons to send an email or make a call in the Zoho CRM App.

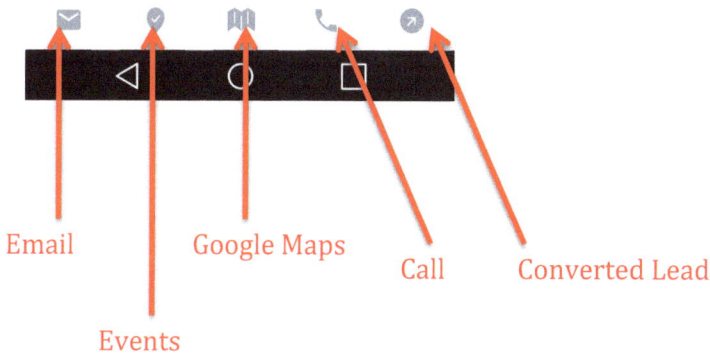

Email

Events

Google Maps

Call

Converted Lead

Settings in the Zoho CRM Mobile App

Scroll down in the menu, in the upper left corner, to see the Setting Menu.

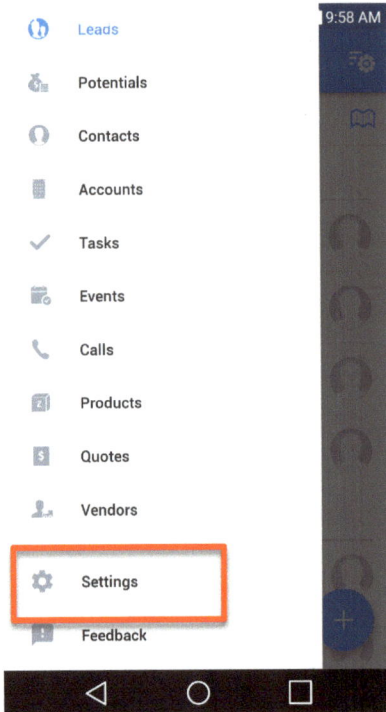

Logging Out of the Zoho CRM Mobile App

There is an icon to log out in the settings in the upper right-hand corner.

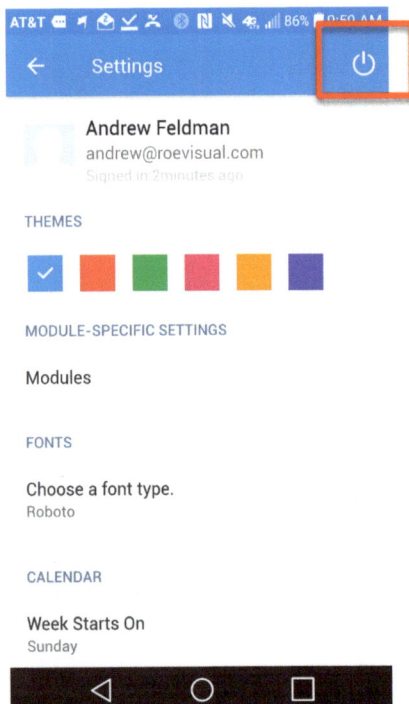

Editing the Smart View Settings in the Mobile Zoho CRM App

Smart View controls how you see only the fields in a record with values in them and whether you have the option to 'See all fields'. If you don't like those settings you can change them here.

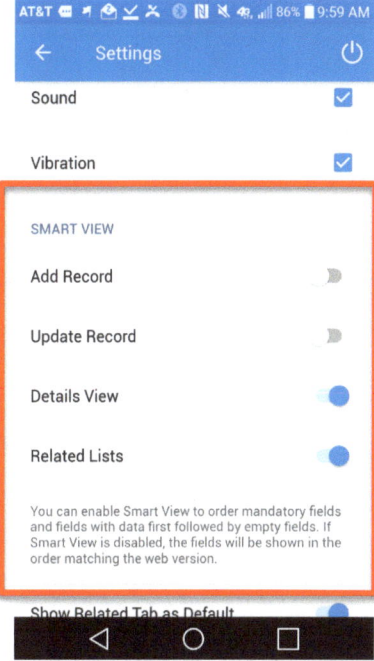

Resetting the App to be Consistent with the Web Version on Zoho CRM Mobile App

If you find that fields are missing in the mobile app to what you have in the web version, you can make the Zoho CRM app reset to the web application by clicking 'Reset App'. This will update the app and make sure it shows all fields in the web app. Make sure, however, that you are not missing fields due to Smart View which hides fields with no values in them until you click 'See all fields'.

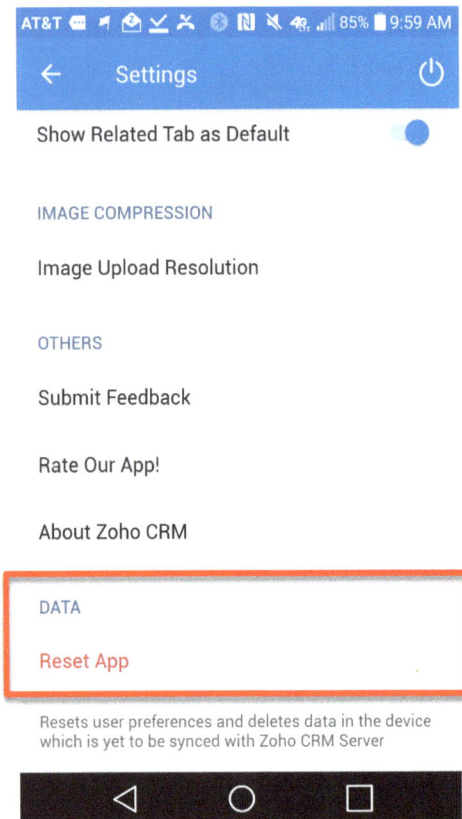

Index

About the Author

Named Wonderpreneur on Medium.com, Zoho One Expert, Financial Planner App developer, Author-Technical Visual Guides

Ebitari Larsen founded DDS in 2009 with the goal, though critical thinking, software technology development and implementation, business analysis and project management, to assist companies with their greatest data management challenges. Coming from a management consulting background she has successfully worked with numerous federal government agencies. She is an award-winning researcher and teacher. Ms. Larsen has assisted prominent establishments to see tangible achievements towards their data management goals including Rutgers University, US Department of Energy (DOE), Comcast Programming Group and numerous small/medium sized enterprises and non-profits.

Zoho Specialties: Zoho One • Zoho CRM • Zoho Books and Invoice • Zoho Sites • Zoho Campaigns • Zoho Creator • Zoho Analytics • Zoho People • Zoho Recruit • Zoho Forms and Survey • Zoho Sign • Zoho Bookings

Functional IT Specialties: Project Management • Business Process Reengineering • Requirement Gathering • Technical Documentation including Videos • Testing • Technical Training

Partners: Zoho, Google, Acxiom, Laserfiche

Some neat tit bits about Ebitari:
* Former avid surfer in Hawaii - now a fair-weather surfer in California :-)
* Former Certified Scientific Diver at University of Hawaii
* Bowdoin College 1994 graduate and ran track one season
* Columbia University Klingenstein Fellow for new teachers 1996 while at Punahou School
* For fun enjoys golf and time in the water (ocean, lake, pool)
* Enjoys spending quality time with her husband, Matthew, and 8-year-old son, Briggs - avid fishermen!
* Likes to help out at church with Children and Youth
* Pronouns she, her, hers

Contact the author at www.dds-llc.com